Editor
Leasha Taggart

Editorial Manager
Karen J. Goldfluss, M.S. Ed.

Editor in Chief
Sharon Coan, M.S. Ed.

Illustrator
Chandler Sinnott

Cover Artist
Jessica Orlando

Creative Director
Elayne Roberts

Art Coordinator
Denice Adorno

Product Manager
Phil Garcia

Imaging
Ralph Olmedo, Jr.

Publishers
Rachelle Cracchiolo, M.S. Ed.
Mary Dupuy Smith, M.S. Ed.

How Add & Subtract

Grade 1

Author

Mary Rosenberg

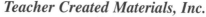

Teacher Created Materials, Inc.
6421 Industry Way
Westminster, CA 92683
www.teachercreated.com

ISBN 1-57690-942-5

©2000 Teacher Created Materials, Inc.
Reprinted, 2000
Made in U.S.A.

Table of Contents

A Note to Teachers and Parents

Welcome to the "How to" math series! You have chosen one of over two dozen books designed to give your children the information and practice they need to acquire important concepts in specific areas of math. The goal of the "How to" math books is to give children an extra boost as they work toward mastery of the math skills established by the National Council of Teachers of Mathematics (NCTM) and outlined in grade-level scope and sequence guidelines. The NCTM standards encourage children to learn basic math concepts and skills and apply them to new situations and to real-world events. The children learn to justify their solutions through the use of pictures, numbers, words, graphs, and diagrams.

The design of this book is intended to allow it to be used by teachers or parents for a variety of purposes and needs. Each of the units contains one or more "How to" pages and two or more practice pages. The "How to" section of each unit precedes the practice pages and provides needed information such as a concept or math rule review, important terms and formulas to remember, or step-by-step guidelines necessary for using the practice pages. While most "How to" pages are written for direct use by the children, in some lower-grade level books these pages are presented as instructional pages or direct lessons to be used by a teacher or parent prior to introducing the practice pages. In this book, the "How to" pages detail the concepts that will be covered in the pages that follow as well as how to teach the concept(s). Many of the "How to" pages also include "learning tips" and "extension ideas." The practice pages review new skills and provide opportunities for the children to apply the newly acquired skills. Each unit is sequential and builds upon the ideas covered in the previous unit(s).

About This Book

How to Add & Subtract: Grade 1 presents a comprehensive overview of addition and subtraction of whole numbers on a level appropriate to students in grade 1. The clear, simple, readable instruction pages for each unit make it easy to introduce and teach basic addition and subtraction to children with little or no background in these concepts.

The use of manipulatives as visualization tools is encouraged. In addition, if children have difficulty with a specific concept or unit within this book, review the material and allow them to redo troublesome pages. Since concept development is sequential, it is not advisable to skip much of the material in the book. It is preferable that children find the work easy and gradually advance to the more difficult concepts at a comfortable pace.

The following skills are introduced in this book:

- adding to 18
- subtracting to 18
- place value (tens and ones)
- problem solving
- probability and statistics

- estimating
- number sense
- communication of math ideas
- developing math reasoning skills

The units in this book are designed to match the suggestions of the National Council of Teachers of Mathematics (NCTM). They strongly support the learning of addition and subtraction and other processes in the context of problem solving and real-world applications. Use every opportunity to have students apply these new skills in classroom situations and at home. This will reinforce the value of the skill as well as the process. This book matches a number of NCTM standards including these main topics and specific features:

Problem Solving

The children develop and apply strategies to solve problems, verify and interpret results, sort and classify objects, and solve word problems.

Communication

The children are able to communicate mathematical solutions through manipulatives, pictures, diagrams, numbers, and words. They are able to relate everyday language to the language and symbols of math. Children also have the opportunities to read, write, discuss, and listen to math ideas.

Reasoning

Children make logical conclusions through interpreting graphs, patterns, and facts. The children are able to explain and justify their math solutions.

Connections

Children are able to apply math concepts and skills to other curricular areas and to the real world.

Number Sense and Numeration

Children learn to count, label, and sort collections as well as learn the basic math operations of adding and subtracting.

Concepts of Whole Number Operations

Children develop an understanding for the operation (add, subtract) by modeling and discussing situations relating math language and the symbols of operation (+, –) to the problem being discussed.

Other Standards

Children work toward **whole number computation** mastery as they model, explain, and develop competency in basic facts, mental computation, and **estimation** techniques.

Children explore **geometry** and develop **spatial sense** by describing models, drawing and classifying shapes, and relating geometric ideas to number and measurement ideas.

Children learn about **statistics** and **probability** by collecting and organizing data into graphs, charts, and tables.

 **How to** •••••••••••••••••••••••••••• **Count**

Learning Notes

In this unit the children work with the concept of one-to-one correspondence. They count from 0 to 10 and are introduced to the concept of *greater than* and *less than*.

Materials

- a number line to 10, made from a sentence strip, a piece of 2" x 18" (5 cm x 46 cm) construction paper, or written across the top of the page

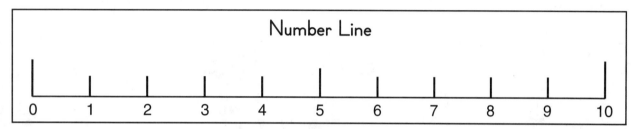

- counters (bottle caps, pennies, counting cubes, paper clips, buttons, craft sticks, beans, etc.)
- pattern blocks

Teaching the Lesson

Before doing the activities, practice counting from 0 to 10 with the children. Ask the children to show you a certain number of fingers or counters or to draw a certain number of stars, happy faces, circles, etc. (For example: "Show me 3 fingers" or "Show me 7 toothpicks.")

Counting to 10 (page 6): Ask the children to count and record the number of ice cream cones in each problem. The children will read the number and make the corresponding number of stars. (The concept of "0" can be difficult for some children to learn.) Have the children look at the boxes where there is "0" and ask them to tell you how many items they see or need to make.

Counting Sides and Corners (page 7): Go over the names of each shape with the children. Have them run their fingers around the sides and corners of each block and describe to you what a "side" is like and what a "corner" is like before doing the activity.

Counting More and Counting Less (page 8): Introduce the concept of *greater than* and *less than* through the use of pictures. Go over the vocabulary with the children and have them show you "more than 5 fingers" or "less than 2 fingers," etc., before beginning the activity.

Extension Idea

Have the children make flashcards using 3" x 5" (8 cm x 13 cm) index cards and stamps or stickers. Place a specific number of stamps or stickers on each card and write the corresponding number on the back. The children can count the items and then check their counting by reading the number on the back. They can also make the corresponding number of items on a separate sheet of paper.

Count the number of items in each problem. Write the number on each line.

1.

2.

3.

4.

5.

_____ _____ _____ _____ _____

6.

7.

8.

9.

10.

_____ _____ _____ _____ _____

Read each number. Make the same number of stars in each box.

2	5	9	0	10
4	1	7	3	6

Now Try This

Write the numbers from 0 to 10.

_____

A square
has 4 equal sides
and 4 corners.

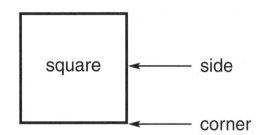

square ⟵ side

⟵ corner

Look at each shape. Count the number of sides and corners. Write the numbers on the lines.

1.

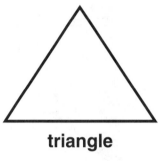

triangle

_____ sides

_____ corners

2.

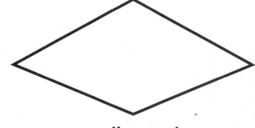

diamond

_____ sides

_____ corners

3.

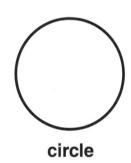

circle

_____ sides

_____ corners

4.

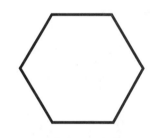

hexagon

_____ sides

_____ corners

5. Make a design by putting several shapes together. (Make the design on a separate sheet of paper.) Count the number of sides and corners in your design and write the numbers on the lines. In the box, write the name of your design.

My design has:

_____ sides

_____ corners

Count the number of items in each group. Write the number of items on the line.
In each box, circle the group that has more items is it.

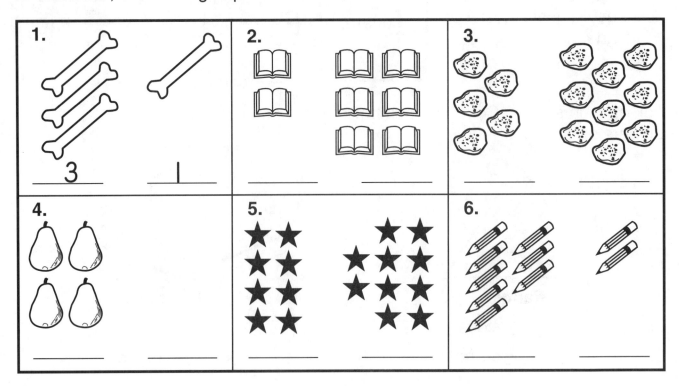

1.		2.	3.

Count the number of items in each group. Write the number of items on the line.
In each box, circle the group that has fewer items in it.

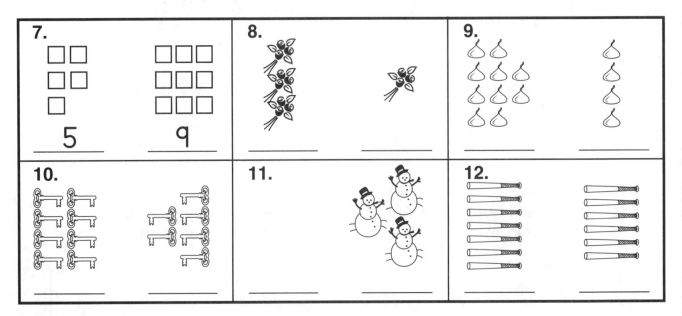

Now Try This: Write the numbers from 10 to 0.

Learning Notes

The practice pages in this unit help children learn to add 2 sets of numbers together to solve an addition problem. The children collect information and make a graph.

Materials

- counters (bottle caps, pennies, paper clips, crayons, craft sticks, multilink cubes, buttons, coins, etc.)

Teaching the Lesson

Before introducing addition to the children, make sure the children have had plenty of experience with counting objects, manipulating objects, putting objects into sets, and counting how many objects there are in all.

When introducing the lesson, use manipulatives (counters, toys, or any kind of objects that can be easily handled) to show the adding process. When collecting data for the graph, remind the children to record each piece of information at the time it is collected!

Learning Tip

Some children have difficulty keeping track of where they are on a math page. A "window" can be made from a 3" x 5" (8 cm x 13 cm) index card. Take the index card and fold it in half. On the fold, cut a rectangle. Open the index card and place the "window" over the math problem on which you are working. This helps the children to quickly and easily locate where they are on a page.

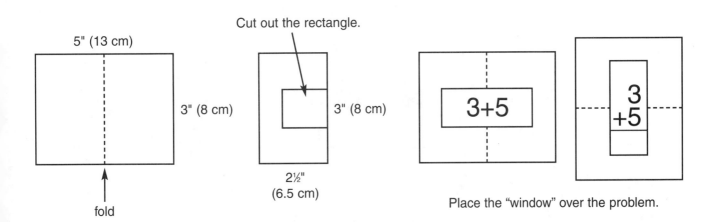

A self-sticking note can also be used as a place marker. Place a brightly colored note under the current problem. As the children complete the problem, they can easily move the note to the next problem.

Extension Tip

Children can make their own math problem using 3" x 5" (8 cm x 13 cm) index cards and stickers or small stamps. On the front side of the index card, the children can stamp out an addition problem, and on the back side, the children can write the answer.

Write the answer to each addition problem.

1.  = _____ rabbits

2.  = _____ cows

3.  = _____ butterflies

Write the addition problem for each set of pictures.

4.  = _____ snails

_____   _____

5.  = _____ books

_____   _____

6.  = _____ birds

_____   _____

Use counters to solve each problem. Write the answer on each line.

1. $0 + 0 =$ _____

6. $2 + 0 =$ _____

2. $0 + 4 =$ _____

7. $5 + 1 =$ _____

3. $1 + 2 =$ _____

8. $0 + 6 =$ _____

4. $1 + 0 =$ _____

9. $4 + 0 =$ _____

5. $4 + 2 =$ _____

10. $0 + 2 =$ _____

Fact Family

A *fact family* is a set of math problems that use the same three numbers. For example, the numbers 1, 3, and 4 can be used to make two addition problems, $1 + 3 = 4$ and $3 + 1 = 4$.

Use the numbers 1, 2, and 3 to make two addition problems.

11.   _____ **+** _____ **=** _____

12.   _____ **+** _____ **=** _____

Ask your family or classmates whether they like cats, dogs, or birds best. Fill in the graph to show the animal that each person chose.

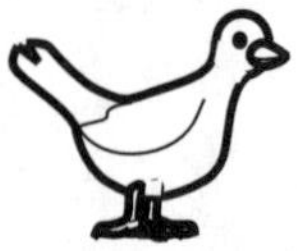 🐦						
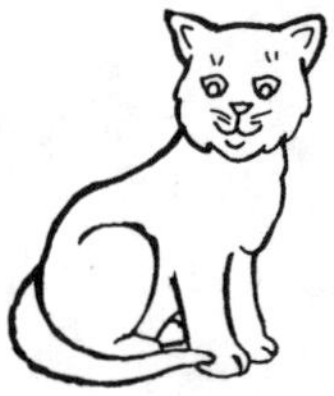 🐱						
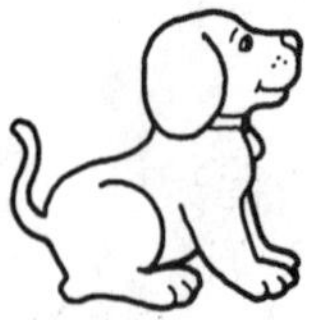 🐶						

1. How many people liked birds best? _____

2. How many people liked cats best? _____

3. How many people liked dogs best? _____

4. Which animal did most people like the best? _____

5. Which animal did the fewest number of people like? _____

Write a sentence about the results.

_____

- -

_____

- -

_____

Learning Notes

In this unit the children use manipulatives to practice adding to 10. They also use their adding skills to solve word problems.

Materials

- counters (beans, teddy bear counters, craft sticks, game pieces, etc.)
- paper cups
- circles (either drawn on a piece of paper or use two jar lids or butter tub lids)

Teaching the Lesson

Adding to 10 (page 14): Model for the children how to put counters representing the first number in the math problem in the first circle and how to put counters representing the second number in the second circle. Ask the children to move the counters to a third circle and count all of the counters with you. Reread the math problem with the answer. Write the answer on the line. For example, in the sentence $3 + 6 = \underline{\quad}$, count aloud "1, 2, 3, 4, 5, 6, 7, 8, 9. $3 + 6 = 9$." Write the answer on the line.

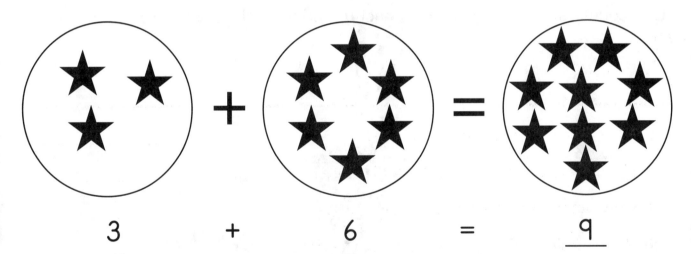

$$3 \qquad + \qquad 6 \qquad = \qquad \underline{9}$$

Adding to 10 and Using the Commutative Property (page 15): Use craft sticks and a cup to introduce this property. Introduce the Commutative Property, which demonstrates that the order of the addends does not change the sum. The same addends (the numbers being added) are used in 2 different math problems. The answer is the same in both cases, only the addends have changed places.

Addition Using Pictures and Sentences (page 16): The children will solve word problems with strawberries and baskets. The children need to draw pictures in order to solve the word problems.

Extension Idea

Cut eleven 3" x 5" (8 cm x 13 cm) index cards in half. Make two sets of cards numbered 0–10. On each card, make the same number of dots, stamps, stars, etc. Mix the cards together and place in one stack. Take the top two cards and add the numbers together.

To extend this activity further, have the child write the math problems down on a piece of paper or in a math journal.

When adding, two sets of number are put together to make one larger set.

For example: 2 + 6 = ___

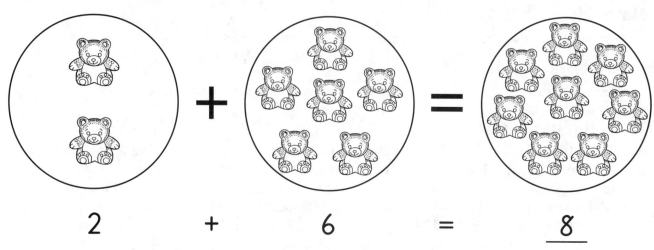

$$2 \qquad + \qquad 6 \qquad = \qquad \underline{8}$$

Use counters and circles to solve each problem. Write the answer on the line.

1. 5 + 2 = ___   7. 3 + 5 = ___

2. 6 + 3 = ___   8. 2 + 6 = ___

3. 8 + 1 = ___   9. 3 + 4 = ___

4. 7 + 2 = ___   10. 5 + 3 = ___

5. 10 + 0 = ___   11. 1 + 9 = ___

6. 5 + 4 = ___   12. 3 + 7 = ___

When adding to 10, craft sticks and a paper cup can be used to solve each problem. For example, to solve 5 + 5 = ___, place 5 craft sticks in the cup. Next, add 5 more craft sticks to the cup. How many sticks are in the cup?

1. 7 + 1 = ___

2. 5 + 0 = ___

3. 6 + 1 = ___

4. 3 + 3 = ___

5. 7 + 0 = ___

6. 1 + 7 = ___

7. 0 + 8 = ___

8. 3 + 6 = ___

9. 3 + 7 = ___

10. 0 + 0 = ___

The Commutative Property occurs when the two addends (the numbers being added together) trade places. For example, in the number sentences 3 + 7 = ___ and 7 + 3 = ___, the "3" and the "7" have traded places, but the answer is still the same.

Write addition problems using the commutative property.

11. 3 + 7 = 10
     7 + 3 = 10

12. 4 + 6 = ___
     ___ + ___ = ___

13. 1 + 9 = ___
     ___ + ___ = ___

14. 2 + 8 = ___
     ___ + ___ = ___

Review

Use counters or craft sticks to solve the problems below. Write the answer to each problem below the line.

15. 5
    + 1
    ___

16. 3
    + 2
    ___

17. 0
    + 4
    ___

18. 1
    + 4
    ___

19. 2
    + 2
    ___

6 strawberries can fit into 1 basket.

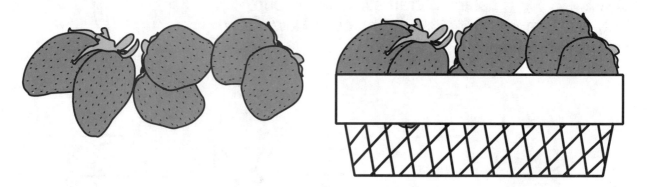

1. How many baskets would you need for 10 strawberries? Use pictures and numbers to show how you solved the problem.

Write a sentence about your answer.

_____

- -

_____

2. How many strawberries could fit into 3 baskets? Use pictures and numbers to show how you solved the problem.

Write a sentence about your answer.

_____

- -

_____

Learning Notes

In this unit children learn to use a number line and practice adding to 18 by counting on it.

Materials

- a number line to 20, made from a sentence strip, a piece of 2" x 18" (5 cm x 45 cm) construction paper, or written across the top of the page

Number Line

```
|  |  |  |  |  |  |  |  |  |  |  |  |  |  |  |  |  |  |  |  |
0  1  2  3  4  5  6  7  8  9 10 11 12 13 14 15 16 17 18 19 20
```

Teaching the Lesson

Using and Counting on a Number Line (pages 18 and 19): Make the number line first. Show the children how to start at a given number and then count to arrive at the answer.

As an example, use the number sentence $9 + 7 =$ ___. Put your finger on 9, then count forward 7 spaces to arrive at the answer 16.

Do several problems with the children until the children feel comfortable using the number line.

Then have children complete pages 18 and 19 using the number line.

Addition Word Problems (page 20): Before introducing the word problems on page 20, practice verbally solving word problems with the children. (Example Problem: I received 8 stars. I earned 2 more stars. How many stars do I have now?)

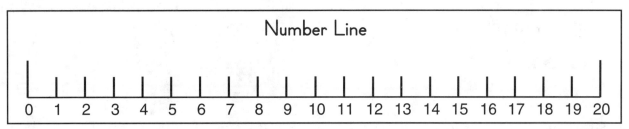

Read the word problems with the children. Point out how the math problem is made from the numbers used in the word problem. Complete the word problems on the page.

Learning Tip

When making a number line, use different colors of sticker dots. Place the sticker dots on the number line and number the dots 1 to 18. The sticker dots provide a visual as well as a tactile cue to the child.

Use the number line to help you solve the word problems.

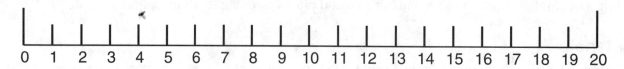

1. Start on 5. Count forward 7. What number are you on?

   _____

2. Start on 4. Count forward 10. What number are you on?

   _____

3. Start on 8. Count forward 10. What number are you on?

   _____

4. Start on 6. Count forward 9. What number are you on?

   _____

5. Start on 9. Count forward 9. What number are you on?

   _____

6. Start on 7. Count forward 0. What number are you on?

   _____

Use the number line to help you solve the addition problems.

7.
$\begin{array}{r} 3 \\ + 9 \\ \hline \end{array}$

8.
$\begin{array}{r} 5 \\ + 8 \\ \hline \end{array}$

9.
$\begin{array}{r} 13 \\ + 3 \\ \hline \end{array}$

10.
$\begin{array}{r} 0 \\ + 11 \\ \hline \end{array}$

11.
$\begin{array}{r} 7 \\ + 7 \\ \hline \end{array}$

12.
$\begin{array}{r} 6 \\ + 10 \\ \hline \end{array}$

13.
$\begin{array}{r} 2 \\ + 12 \\ \hline \end{array}$

14.
$\begin{array}{r} 4 \\ + 13 \\ \hline \end{array}$

15.
$\begin{array}{r} 10 \\ + 1 \\ \hline \end{array}$

16.
$\begin{array}{r} 17 \\ + 1 \\ \hline \end{array}$

17.
$\begin{array}{r} 18 \\ + 0 \\ \hline \end{array}$

18.
$\begin{array}{r} 9 \\ + 7 \\ \hline \end{array}$

Counting on a number line is a fast way to arrive at an answer.

For example, in the number sentence, 9 + 7 = ___, start on 9, and then count forward 7 more spaces to arrive at the answer.

Use the number line to *count on* from a given number.

1.	2.	3.	4.	5.
5 + 11	14 + 0	13 + 1	11 + 4	12 + 5

6.	7.	8.	9.	10.
10 + 2	3 + 12	7 + 10	5 + 13	8 + 6

11.	12.	13.	14.	15.
7 + 9	6 + 9	15 + 3	14 + 3	2 + 11

16.	17.	18.	19.	20.
2 + 13	4 + 10	4 + 12	5 + 10	1 + 14

Mental Math

Add the numbers. Write the answer on the line.

21. $1 + 1 + 1 = $ ___

22. $1 + 2 + 3 = $ ___

23. $3 + 4 + 5 = $ ___

24. $5 + 1 + 4 = $ ___

25. $2 + 4 + 6 = $ ___

26. $3 + 3 + 3 = $ ___

Read each word problem and write the numbers to make an addition problem. Write the answer below the addition word problem.

1. Kevin has 7 marbles. His dad gives him 5 more. How many marbles does Kevin have now? _____ marbles	——— + ——— 	2. Julia found 9 pennies. Then she found 8 more. How many pennies does Julia have now? _____ pennies	——— + ———

1. Kevin has 7 marbles. His dad gives him 5 more. How many marbles does Kevin have now?

———
+ ———

_____ marbles

2. Julia found 9 pennies. Then she found 8 more. How many pennies does Julia have now?

———
+ ———

_____ pennies

3. Rachel made 10 place mats. Her dad gave her 4 more. How many place mats does Rachel have now?

———
+ ———

_____ place mats

4. Jacob has 6 toy dinosaurs. He bought 6 more. How many toy dinosaurs does Jacob have now?

———
+ ———

_____ dinosaurs

5. Ben made 3 cookies. Then he made 11 more cookies. How many cookies does Ben have now?

———
+ ———

_____ cookies

6. Beth has 2 teddy bears. Her grandma gave her 12 more. How many teddy bears does Beth have now?

———
+ ———

_____ teddy bears

7. Clare planted 8 carrot seeds and 9 tomato seeds. How many seeds did Clare plant in all?

———
+ ———

_____ seeds

8. Kim has 14 pencils. She did not buy anymore. How many pencils does Kim have?

———
+ ———

_____ pencils

9. Write and solve your own addition word problem.

_____

- -

_____   ———

_____   + ———

- -

Learning Notes

In this unit the children will use manipulatives to subtract one number from another.

Materials

- counters (beans, bottle caps, pennies, paper clips, crayons, toothpicks, multilink cubes, etc.)

Teaching the Lesson

Subtracting to 6 (pages 22–24): Before beginning the practice pages, have the children practice subtracting objects to verbally given word problems. As an example, introduce the following problem: There are 6 frogs and 2 hop away. How many frogs are left?

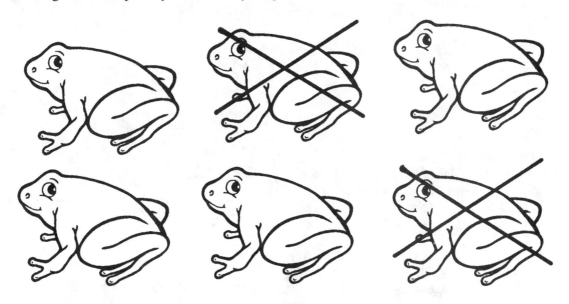

The children will show 6 counters, subtract (cross out) 2, and count the remaining counters to arrive at the answer.

When working on the activities, model how to cross off (subtract) the correct number of pictures.

Learning Tip

Children can learn to check their own answers by adding. For example, in the number sentence, 9 – 6 = 3, check the answer by adding 3 (the answer) and 6 (the minuend) to solve for 9.

Checking answers also solidifies the concept of fact families as well as shows the relationship between adding and subtracting.

Extension Idea

Children can make their own math problems using 3" x 5" (8 cm x 13 cm) index cards and stickers or small stamps. On the front side of the index card, the children can stamp out a subtraction problem and then cross off a selected number of the pictures. On the back of the index card, the children can write the answer to the problem.

Subtraction means to *take away* a certain amount from another number.

For example, in the number sentence,
$6 - 2 =$ _____, start with 6 objects. Cross
off (take away) 2. The remaining number of
items is the answer.

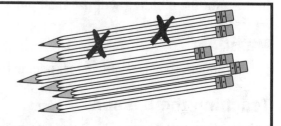

Read each problem. Subtract (cross off) the correct number of pictures. Write
the answer on the line.

1.

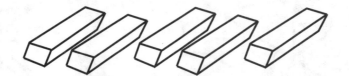

$5 - 2 =$ _____

2.

$6 - 4 =$ _____

3.

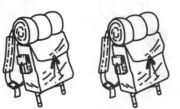

$2 - 2 =$ _____

4.

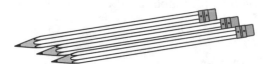

$3 - 0 =$ _____

5.

$6 - 5 =$ _____

6.

$4 - 3 =$ _____

5 ▶ Practice •••••••••••• Writing the Math Problem

When writing subtraction problems, always start with the total number of items. Then write the number that is being subtracted and the answer.

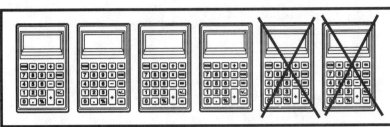

Example: There were 6 calculators but 2 were taken away. The math problem is 6 – 2 = 4.

1.

6 – _____ = _____

2.

4 – _____ = _____

3.

5 – _____ = _____

4.

3 – _____ = _____

5.

2 – _____ = _____

6.

6 – _____ = _____

7.

4 – _____ = _____

© *Teacher Created Materials, Inc*

Use counters to solve each problem. Write the answer on the line.

1. 6 – 6 = ____

2. 6 – 4 = ____

3. 5 – 1 = ____

4. 4 – 2 = ____

5. 2 – 0 = ____

6. 4 – 4 = ____

7. 2 – 1 = ____

8. 4 – 0 = ____

9. 5 – 5 = ____

10. 6 – 2 = ____

11. 3 – 0 = ____

12. 3 – 3 = ____

Fact Family

A *fact family* is a set of math problems that use the same three numbers. For example, the numbers 2, 3, and 5 can be used to make two subtraction problems, 5 – 3 = 2 and 5 – 2 = 3. Remember, when subtracting, always start with the larger number.

Use the numbers 1, 2, and 3 to make two subtraction problems.

13. _____ – _____ = _____

14. _____ – _____ = _____

Learning Notes

In this unit the children will learn to subtract to 10 through the use of manipulatives.

Materials

- counters (bottle caps, pennies, paper clips, crayons, beans, etc.)
- craft sticks
- paper cups
- circle (the lid to a butter tub or a jar or a circle drawn on a piece of paper)

Teaching the Lesson

Subtracting to 10 (page 26): Model for the children how to put counters representing the largest number in the circle and take away the number representing the minuend (the number being subtracted). Ask the children to count the remaining counters (those still in the circle) with you. Write the answer on the line.

Subtracting to 10 with Craft Sticks (page 27): Use the same procedure outlined above but with craft sticks.

Using the Subtraction Fact Family (page 28): Review the fact family with the children. Practice several examples together before beginning the practice page.

Extension Idea

Cut eleven 3" x 5" (8 cm x 13 cm) cards in half. Make two sets of cards numbered 0–10. Shuffle the cards together and place them in one stack facedown. Take the top two cards and use them to make a subtraction problem. Remind the children that when subtracting, always start with the larger number first.

When subtracting, one number is being *taken away* from a larger number.

Example:

8 – 5	**=**	**3**

Complete each subtraction problem.

1. 10 – 8 = ___

2. 8 – 6 = ___

3. 7 – 2 = ___

4. 6 – 1 = ___

5. 9 – 8 = ___

6. 8 – 4 = ___

7. 10 – 10 = ___

8. 6 – 5 = ___

9. 9 – 0 = ___

10. 7 – 5 = ___

11. 10 – 5 = ___

12. 8 – 1 = ___

When subtracting to 10, craft sticks and a paper cup can be used to solve each problem.

For example: 10 − 5 = ___

Put 10 craft sticks in the cup, remove 5, and count how many craft sticks are left.

Use craft sticks to solve the subtraction problems.

Write the answer on the line.

1. 10 − 7 = ___

2. 5 − 4 = ___

3. 7 − 7 = ___

4. 10 − 3 = ___

5. 8 − 3 = ___

6. 6 − 3 = ___

7. 10 − 4 = ___

8. 9 − 4 = ___

9. 7 − 3 = ___

10. 6 − 0 = ___

11. 9 − 1 = ___

12. 3 − 3 = ___

13. 4 − 1 = ___

14. 10 − 2 = ___

15. 8 − 5 = ___

16. 7 − 1 = ___

17. 6 − 4 = ___

18. 7 − 0 = ___

19. 9 − 5 = ___

20. 3 − 2 = ___

21. On the back of this paper, make a list of all of the math problems that had 3 as the answer.

Write the other subtraction problem that belongs in the same fact family. The first one has been done for you.

1. $10 - 4 = 6$ $10 - 6 = 4$	5. $6 - 0 =$ ___ ___ $-$ ___ $=$ ___
2. $8 - 6 =$ ___ ___ $-$ ___ $=$ ___	6. $4 - 3 =$ ___ ___ $-$ ___ $=$ ___
3. $9 - 2 =$ ___ ___ $-$ ___ $=$ ___	7. $10 - 1 =$ ___ ___ $-$ ___ $=$ ___
4. $7 - 4 =$ ___ ___ $-$ ___ $=$ ___	8. $5 - 2 =$ ___ ___ $-$ ___ $=$ ___

Review

Use counters or craft sticks to solve the problems below. Write the answer to each problem below the line.

9. $\begin{array}{r} 10 \\ -\ 0 \\ \hline \end{array}$
10. $\begin{array}{r} 9 \\ -\ 6 \\ \hline \end{array}$
11. $\begin{array}{r} 8 \\ -\ 2 \\ \hline \end{array}$
12. $\begin{array}{r} 5 \\ -\ 5 \\ \hline \end{array}$
13. $\begin{array}{r} 10 \\ -\ 6 \\ \hline \end{array}$
14. $\begin{array}{r} 7 \\ -\ 6 \\ \hline \end{array}$

Learning Notes

In this unit the children will learn to use a number line for subtraction problems. They will also practice subtracting to 18.

Materials

- a number line to 20 made from a sentence strip, a piece of 2" x 18" (5 cm x 46 cm) construction paper, or written across the top of the page.

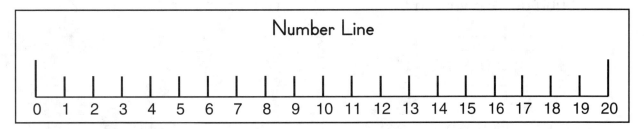

Teaching the Lesson

Using a Number Line and **Counting Backwards (pages 30–32):** Make the number line first. Show the children how to start at a given number and then count backwards for subtraction problems.

Verbally, give the children a subtraction word problem. For example: I made 12 cookies. I ate 9 of them. How many cookies do I have left?

The children start on 12 and then count backwards (take away) 9 to arrive at the answer (3).

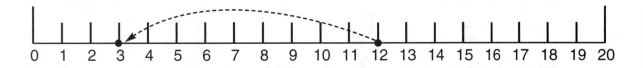

Do this several times until the children feel comfortable working with a number line.

Learning Tip

When making a number line, use different colors of sticker dots. Place the sticker dots on the number line and number the dots 1 to 18. The sticker dots provide a visual as well as a tactile cue to the child.

Use the number line to help you solve the word problems.

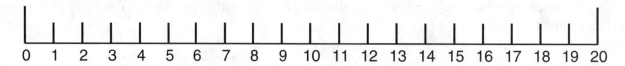

1. Start on 15. Count backward 9.
   What number are you on?

   _____

4. Start on 16. Count backward 7.
   What number are you on?

   _____

2. Start on 12. Count backward 3.
   What number are you on?

   _____

5. Start on 14. Count backward 5.
   What number are you on?

   _____

3. Start on 18. Count backward 0.
   What number are you on?

   _____

6. Start on 17. Count backward 16.
   What number are you on?

   _____

Use the number line to help you solve the addition problems.

7. $\begin{array}{r} 16 \\ -\ 9 \\ \hline \end{array}$

8. $\begin{array}{r} 15 \\ -\ 5 \\ \hline \end{array}$

9. $\begin{array}{r} 18 \\ -14 \\ \hline \end{array}$

10. $\begin{array}{r} 13 \\ -\ 9 \\ \hline \end{array}$

11. $\begin{array}{r} 12 \\ -\ 6 \\ \hline \end{array}$

12. $\begin{array}{r} 11 \\ -\ 8 \\ \hline \end{array}$

13. $\begin{array}{r} 18 \\ -12 \\ \hline \end{array}$

14. $\begin{array}{r} 15 \\ -\ 8 \\ \hline \end{array}$

15. $\begin{array}{r} 11 \\ -\ 1 \\ \hline \end{array}$

16. $\begin{array}{r} 18 \\ -\ 8 \\ \hline \end{array}$

17. $\begin{array}{r} 12 \\ -11 \\ \hline \end{array}$

18. $\begin{array}{r} 11 \\ -10 \\ \hline \end{array}$

Counting backwards is a fast way to arrive at an answer.

For example, in the number sentence, $17 - 7 =$ ___, start on 17 and then *count backward* 7 spaces to arrive at the answer.

Use the number line to *count backwards*. Write the answer below the line.

1.
$$10$$
$$-\ 7$$

2.
$$15$$
$$-\ 10$$

3.
$$10$$
$$-\ 6$$

4.
$$14$$
$$-\ 12$$

5.
$$17$$
$$-\ 13$$

6.
$$12$$
$$-\ 8$$

7.
$$16$$
$$-\ 2$$

8.
$$13$$
$$-\ 5$$

9.
$$14$$
$$-\ 10$$

10.
$$15$$
$$-\ 7$$

11.
$$18$$
$$-\ 12$$

12.
$$11$$
$$-\ 6$$

13.
$$18$$
$$-\ 16$$

14.
$$14$$
$$-\ 8$$

15.
$$13$$
$$-\ 4$$

16.
$$18$$
$$-\ 3$$

17.
$$12$$
$$-\ 4$$

18.
$$16$$
$$-\ 7$$

19.
$$13$$
$$-\ 3$$

20.
$$16$$
$$-\ 5$$

Mental Math

Subtract the numbers. Use the number line to help you. Write the answer on the line.

21. $15 - 9 - 5 =$ ___

22. $14 - 3 - 0 =$ ___

23. $12 - 4 - 6 =$ ___

24. $17 - 0 - 17 =$ ___

25. $18 - 9 - 4 =$ ___

26. $16 - 8 - 8 =$ ___

Read each word problem and write the numbers to make a subtraction problem. Write the answer below each problem.

1. Janet had 17 books. She gave 9 of the books to her sister. How many books does Janet have left? ———— – ———— ———— books	2. Eric had 18 comic books. 9 of the comic books got wet. How many comic books are not wet? ———— – ———— ———— comic books
3. David had 12 race cars. He gave 4 of the cars to his friend. How many cars does David have left? ———— – ———— ———— cars	4. Jackie had 16 basketballs. 9 of them are flat. How many basketballs still bounce? ———— – ———— ———— basketballs
5. Flora had 15 donuts. She ate 3 of them. How many donuts does Flora have left? ———— – ———— ———— donuts	6. Wilbur had 13 spiders. 1 of them ran away. How many spiders does Wilbur have left? ———— – ———— ———— spiders
7. Zachary had 14 yo-yos. 7 of the yo-yos broke. How many yo-yos does Zachary have left? ———— – ———— ———— yo-yos	8. Yolanda had 10 hats. Her dog ate 8 of them. How many hats does Yolanda have left? ———— – ———— ———— hats

9. Write your own subtraction word problem.

_____

- ————

_____   –   ————

- -

Learning Notes

In this unit the children will be able to apply the strategies they have learned to various addition and subtraction problems to 18.

Materials

- a number line
- counters (beans, craft sticks, coins, paper clips, multilink cubes, etc.)
- circles (made from butter tub or jar lids or circles drawn on a piece of paper)
- craft sticks
- paper cups

Teaching the Lesson

Adding and Subtracting (page 34): Review with the children the different strategies that have been learned for adding and subtracting.

Verbally give the children several addition and subtraction word problems to solve using the different strategies that have been learned so far.

Go over the practice page with the children and discuss various ways of solving the problems. Let each child choose a way to solve the addition and subtraction problems.

Learning Tip

Every child is different and learns differently. Some children like to use fingers to count, others prefer to use number lines, to draw pictures of the problem, or to use counters to solve the problem. Allow each child to use what works best for him or her.

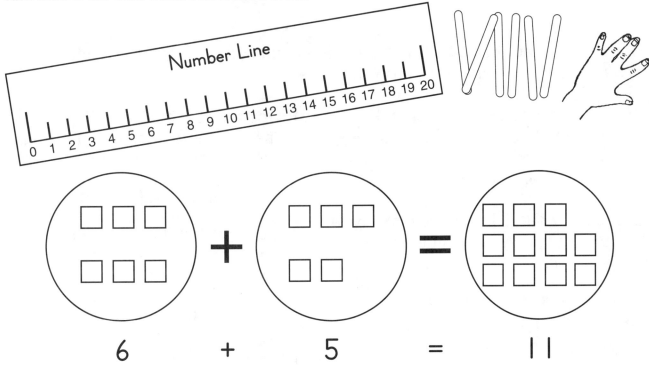

Add or subtract to solve each problem. Pay careful attention to the (+) addition signs and (–) subtraction signs.

1. 10 − 9

2. 6 + 3

3. 8 + 2

4. 6 − 4

5. 5 − 5

6. 8 − 1

7. 7 + 0

8. 4 + 4

9. 0 − 0

10. 9 + 8

11. 3 + 5

12. 2 + 7

13. 2 − 2

14. 10 − 0

15. 5 + 5

16. 7 − 6

17. 9 − 9

18. 3 − 3

Read each word problem. Circle the math operation (add or subtract) that needs to be done to solve the word problem.

19. Melinda had $9. She spent $6 at the pet store. How much money does Melinda have left?

   add   subtract

20. Jeremy had 8 fish in his tank. Now there are only 3 fish in the tank. How many fish are not in the tank?

   add   subtract

21. Bob made 5 baskets. Then he made 6 more baskets. How many baskets did Bob make?

   add   subtract

22. Carrie swam 3 miles this morning and 4 miles this afternoon. How many miles did Carrie swim in all?

   add   subtract

Use the number line to help you solve the word problems.

0 1 2 3 4 5 6 7 8 9 10 11 12 13 14 15 16 17 18 19 20

1. Start on 7. Count forward 3. Count backward 4. What number are you on?

_____

2. Start on 12. Count forward 4. Count backward 0. What number are you on?

_____

3. Start on 10. Count backward 9. Count forward 5. What number are you on?

_____

4. Start on 11. Count backward 4. Count forward 7. What number are you on?

_____

5. Start on 15. Count backward 8. Count forward 6. What number are you on?

_____

6. Start on 4. Count forward 9. Count backward 1. What number are you on?

_____

7. Start on 5. Count backward 1. Count forward 7. What number are you on?

_____

8. Start on 2. Count forward 14. Count backward 10. What number are you on?

_____

9. Start on 1. Count forward 5. Count backward 2. What number are you on?

_____

10. Start on 13. Count backward 11. Count forward 8. What number are you on?

_____

8 ▶ Practice ••••••••••••••••••• The Fact Families

Using each family of numbers, make two addition problems and two subtraction problems.

1.   2, 13, 15
 	
      2 + 13   = 15
    13 + 12 = 15
    15 – 2   = 13
    15 – 13 = 2
	2. 7, 8, 15
	
	 ___ + ___ = ___
	 ___ + ___ = ___
	 ___ – ___ = ___
	 ___ – ___ = ___

3. 6, 9, 15 ___ + ___ = ___ ___ + ___ = ___ ___ – ___ = ___ ___ – ___ = ___	**4.** 4, 11, 15 ___ + ___ = ___ ___ + ___ = ___ ___ – ___ = ___ ___ – ___ = ___
5. 3, 12, 15 ___ + ___ = ___ ___ + ___ = ___ ___ – ___ = ___ ___ – ___ = ___	**6.** 5, 10, 15 ___ + ___ = ___ ___ + ___ = ___ ___ – ___ = ___ ___ – ___ = ___

7. What did you notice about these fact families?

_____

- -

_____

Challenge
On the back make a new *fact family*. Pick three numbers that belong in the same family and make two addition problems and two subtraction problems.

Learning Notes

In this unit the children will learn about place value (tens and ones) and how to add and subtract two-digit numbers without regrouping.

Materials

- beans
- craft sticks
- glue
- plastic storage bags
- 3" x 5" (8 cm x 13 cm) index cards

Before Teaching the Lesson

Have the children make bean sticks. Give each child 99 beans and 9 craft sticks. Have the children glue exactly 10 beans to each craft stick. Leave the remaining 9 beans loose. The craft sticks will represent tens. The loose beans will represent ones. Store the bean sticks and loose beans in a plastic storage bag.

Introduce the concept of tens and ones. Have the children make various one- and two-digit numbers using the bean sticks and beans.

For example, ask children to show you 9, show you 24, and show you 36.

Make a tens and ones mat for each child. Have the children practice adding and subtracting two-digit numbers using the beans, bean sticks, and tens and ones mats. Emphasize to the children the importance of starting on the ones side when adding or subtracting.

tens	ones

Teaching the Lesson

Using Tens and Ones (page 38): Go over the practice page that introduces the concept of tens and ones. Discuss how the single crayons stand for one and the box of crayons stands for ten. Model how to count and record the ones and tens.

Two-Digit Addition and Subtraction (page 39): Give each child a small self-sticking note to use when adding or subtracting two-digit numbers. Model how to use the sticky note to cover up the tens column when adding the ones together and remove the self-sticking note when adding the tens column.

Guessing the Operation (page 40): Have the children number the index cards 0–99. Two or three children can share the same deck of cards. Remind the children that when adding or subtracting, they must start on the ones side. Also, remind them that in order to subtract, the number on the first card must be larger than the number on the second card.

Each box of crayons holds 10 crayons or one set of ten. Each single crayon is one. Count each set of tens and ones. Write the number of tens and ones in each box. Write the total number on the line below.

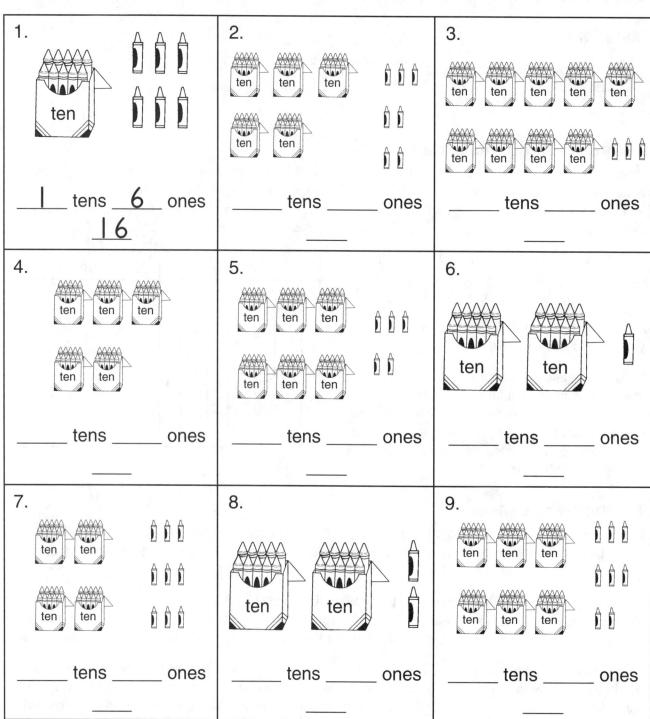

1. ___1___ tens ___6___ ones

___16___

2. _____ tens _____ ones

_____

3. _____ tens _____ ones

_____

4. _____ tens _____ ones

_____

5. _____ tens _____ ones

_____

6. _____ tens _____ ones

_____

7. _____ tens _____ ones

_____

8. _____ tens _____ ones

_____

9. _____ tens _____ ones

_____

10. Write the answers from each problem in order from smallest to greatest.

_____ , _____ , _____ , _____ , _____ , _____ , _____ , _____ , _____

Add the following two-digit numbers (without regrouping).

1.
```
  59
+ 10
```

2.
```
  20
+ 30
```

3.
```
  43
+ 16
```

4.
```
  61
+ 27
```

5.
```
  77
+ 12
```

6.
```
  80
+ 15
```

7.
```
  80
+ 10
```

8.
```
  25
+ 24
```

9.
```
  33
+ 33
```

10.
```
  18
+ 40
```

11.
```
  45
+ 50
```

12.
```
  61
+ 37
```

13.
```
  50
+ 30
```

14.
```
  80
+ 17
```

15.
```
  14
+ 14
```

16.
```
  29
+ 70
```

17.
```
  36
+ 63
```

18.
```
  50
+ 10
```

Subtract the following two-digit numbers (without regrouping).

19.
```
  87
- 25
```

20.
```
  81
- 60
```

21.
```
  35
- 14
```

22.
```
  26
- 16
```

23.
```
  40
- 10
```

24.
```
  12
- 11
```

25.
```
  90
- 90
```

26.
```
  52
- 21
```

27.
```
  47
- 42
```

28.
```
  19
- 14
```

29.
```
  64
- 30
```

30.
```
  95
- 65
```

31.
```
  38
- 17
```

32.
```
  45
- 30
```

33.
```
  96
- 56
```

34.
```
  82
- 20
```

35.
```
  73
- 41
```

36.
```
  18
- 18
```

Use 100 3" x 5" (8 cm x 13 cm) index cards. Number the cards 0–99. Shuffle the cards and place in a stack facedown. Now you are ready to play the game!

- Take the first card and place at the top of the desk.

- Take the second card and place it below the first card.

- Decide whether you are going to add the two cards together or subtract the second card from the first card.

- Record the problems in the boxes below.

1.	2.	3.
4.	5.	6.
7.	8.	9.

Challenge

Put the answers in order from smallest to greatest.

_____, _____, _____, _____, _____, _____, _____, _____, _____

The Ice Cream Shop sells three different flavors of ice cream. The three flavors are chocolate, vanilla, and strawberry. If you order a three-scoop cone with one scoop of each flavor, how many different ways could the scoops of ice cream be arranged? Draw pictures of the different arrangements of ice cream.

Write a sentence about your findings.

_____

- -

_____

- -

_____

·························· **Penny Toss**

If you tossed a penny on the floor 10 times, do you think the penny would land with mostly *heads* up or *tails* up?

I think the penny would land mostly _____ up.

Toss the penny 10 times. After each toss, color in the box to show how the penny landed.

heads							
tails							

The penny landed more times on _____ than on _____.

Was your prediction right? _____

If you tossed the penny on the floor 10 more times, do you think you would have the same results? _____

Toss the penny 10 more times. After each toss, color in a box to show how the penny landed.

heads							
tails							

This time the penny landed more times on _____ than on _____.

Was your second prediction right? _____

Write a sentence telling about your results.

_____

_____

_____

Fill a small cup with beans and make a guess (estimate) about how many beans are in the cup. Write the guess down.

Use an empty egg carton to count the beans. Put 10 beans in each egg space. How many sets of "ten" are there? How many "ones" are there? Write down the actual number of beans that were in the cup.

Do the above steps 2 more times.

Estimate #1

I think there are _____ beans in the cup.

There were actually _____ tens and _____ ones in the cup.

_____ + _____ = _____

Estimate #2

I think there are _____ beans in the cup.

There were actually _____ tens and _____ ones in the cup.

_____ + _____ = _____

Estimate #3

I think there are _____ beans in the cup.

There were actually _____ tens and _____ ones in the cup.

_____ + _____ = _____

Write a sentence telling about your findings.

_____

_____

_____

_____

•••••How Many Outfits Can You Make?

Using 3 shirts and 3 pairs of pants, how many different outfits can you make?
Record the different outfits below.

red   orange   yellow   green   blue   purple

| color of shirt + color of pants = _____ outfit |

_____ + _____ = _____ outfit

_____ + _____ = _____ outfit

_____ + _____ = _____ outfit

_____ + _____ = _____ outfit

_____ + _____ = _____ outfit

_____ + _____ = _____ outfit

_____ + _____ = _____ outfit

_____ + _____ = _____ outfit

_____ + _____ = _____ outfit

How many different outfits did you make?_____

Learning Notes

Use the information below to guide children as they complete practice page 46.

Materials

- computer with paint software
- printer
- page 46
- multilink cubes (optional)

Before Using the Computer

- Review the basics of addition, reminding children that addition means putting together. Multilink cubes can be used as manipulatives to review addition.
- Create the paint file as shown on page 47.
- Children should be familiar with drag and drop in the paint program. They should also know how to use the text tool in the paint program and the print function.

At the Computer

1. Begin by displaying the addition file in the paint program on the monitor.

2. Tell children that they are going to use blocks on the screen to find the answers to addition problems.

3. Show children the first problem and the groups of blocks beside the problem.

4. Demonstrate to children how to drag the blocks into one group.

5. Ask a child to count the number of blocks and record the answer on the computer screen.

6. Show children the next problem and the groups of blocks beside the problem.

7. Have the children record the answer on the screen.

8. Give children the opportunity to solve the problems on the computer.

9. Remind children to close the file without saving the changes. This restores the file to its original state.

Extensions

- Children can solve addition problems using three numbers.
- Children can create addition word problems for others to solve.

Paint File

Name _____

3
$+ 4$ ☆ ☆ ☆ / ☆ ☆ ☆ ☆

4
$+ 6$ ○ ○ ○ ○ / □ □ □ □ □ □

8
$+ 5$ △ △ △ △ △ △ △ △ / D D D D D

7
$+ 6$ □ □ □ □ □ □ □ / ☆ ☆ ☆ ☆ ☆ ☆

9
$+ 2$ ○ ○ ○ ○ ○ ○ ○ ○ ○ / □ □

page 6
1. 4
2. 8
3. 0
4. 10
5. 5
6. 6
7. 3
8. 9
9. 1
10. 7

Now Try This: 0, 1, 2, 3, 4, 5, 6, 7, 8, 9, 10

page 7
1. 3 sides, 3 corners
2. 4 sides, 4 corners
3. 0 sides, 0 corners
4. 6 sides, 6 corners
5. Answers will vary.

page 8
1. already done
2. 2, 6
3. 5, 9
4. 4, 0
5. 8, 10
6. 8, 2
7. already done
8. 3, ①
9. 10, ④
10. 8, ⑥
11. ⓪, 3
12. 7, ⑥

Now Try This: 10, 9, 8, 7, 6, 5, 4, 3, 2, 1, 0

page 10
1. 5
2. 7
3. 6
4. 1 + 3 = 4
5. 4 + 1 = 5
6. 5 + 1 = 6

page 11
1. 0
2. 4
3. 3
4. 1
5. 6
6. 2
7. 6
8. 6
9. 4
10. 2

11 and 12. 1 + 2 = 3
                     2 + 1 = 3

page 12
Answers will vary.

page 14
1. 7
2. 9
3. 9
4. 9
5. 10
6. 9
7. 8
8. 8
9. 7
10. 8
11. 10
12. 10

page 15
1. 8
2. 5
3. 7
4. 6
5. 7
6. 8
7. 8
8. 9
9. 10
10. 0
11. already done
12. 4 + 6 = 10
       6 + 4 = 10
13. 1 + 9 = 10
       9 + 1 = 10
14. 2 + 8 = 10
       8 + 2 = 10
15. 6
16. 5
17. 4
18. 5
19. 4

page 16
1. 2 baskets, sentences will vary.
2. 18 strawberries, sentences will vary.

page 18
1. 12
2. 14
3. 18
4. 15
5. 18
6. 7

7. 12
8. 13
9. 16
10. 11
11. 14
12. 16
13. 14
14. 17
15. 11
16. 18
17. 18
18. 16

page 19
1. 16
2. 14
3. 14
4. 15
5. 17
6. 12
7. 15
8. 17
9. 18
10. 14
11. 16
12. 15
13. 18
14. 17
15. 13
16. 15
17. 14
18. 16
19. 15
20. 15
21. 3
22. 6
23. 12
24. 10
25. 12
26. 9

page 20
1. 7 + 5, 12
2. 9 + 8, 17
3. 10 + 4, 14
4. 6 + 6, 12
5. 3 + 11, 14
6. 2 + 12, 14
7. 8 + 9, 17
8. 14 + 0, 14
9. Answers will vary.

page 22
1. 3   4. 3
2. 2   5. 1
3. 0   6. 1

Page 23
1. 6 − 1 = 5
2. 4 − 1 = 3
3. 5 − 3 = 2
4. 3 − 2 = 1
5. 2 − 0 = 2
6. 6 − 2 = 4
7. 4 − 4 = 0

page 24
1. 0
2. 2
3. 4
4. 2
5. 2
6. 0
7. 1
8. 4
9. 0
10. 4
11. 3
12. 0
13. and 14. 3 − 2 = 1,
                      3 − 1 = 2

page 26
1. 2   7. 0
2. 2   8. 1
3. 5   9. 9
4. 5   10. 2
5. 1   11. 5
6. 4   12. 7

page 27
1. 3
2. 1
3. 0
4. 7
5. 5
6. 3
7. 6
8. 5
9. 4
10. 6
11. 8
12. 0
13. 3
14. 8
15. 3
16. 6
17. 2
18. 7
19. 4
20. 1
21. 10 − 7 = 3, 6 − 3 = 3,
       4 − 1 = 3, 8 − 5 = 3

page 28

1. already done
2. 8 – 6 = 2, 8 – 2 = 6
3. 9 – 2 = 7, 9 – 7 = 2
4. 7 – 4 = 3, 7 – 3 = 4
5. 6 – 0 = 6, 6 – 6 = 0
6. 4 – 3 = 1, 4 – 1 = 3
7. 10 – 1 = 9, 10 – 9 = 1
8. 5 – 2 = 3, 5 – 3 = 2
9. 10
10. 3
11. 6
12. 0
13. 4
14. 1

page 30

1. 6
2. 9
3. 18
4. 9
5. 9
6. 1
7. 7
8. 10
9. 4
10. 4
11. 6
12. 3
13. 6
14. 7
15. 10
16. 10
17. 1
18. 1

page 31

1. 3
2. 5
3. 4
4. 2
5. 4
6. 4
7. 14
8. 8
9. 4
10. 8
11. 6
12. 5
13. 2
14. 6
15. 9
16. 15
17. 8

18. 9
19. 10
20. 11
21. 1
22. 11
23. 2
24. 0
25. 5
26. 0

page 32

1. 8
2. 9
3. 8
4. 7
5. 12
6. 12
7. 7
8. 2
9. Answers will vary.

page 34

1. 1
2. 9
3. 10
4. 2
5. 0
6. 7
7. 7
8. 8
9. 0
10. 17
11. 8
12. 9
13. 0
14. 10
15. 10
16. 1
17. 0
18. 0
19. subtract
20. subtract
21. add
22. add

page 35

1. 6
2. 16
3. 6
4. 14
5. 13
6. 12
7. 11
8. 6
9. 4
10. 10

page 36

1. already done
2. 7 + 8 = 15
   8 + 7 = 15
   15 – 7 = 8
   15 – 8 = 7
3. 6 + 9 = 15
   9 + 6 = 15
   15 – 6 = 9
   15 – 9 = 6
4. 4 + 11 = 15
   11 + 4 = 15
   15 – 4 = 11
   15 – 11 = 4
5. 3 + 12 = 15
   12 + 3 = 15
   15 – 3 = 12
   15 – 12 = 3
6. 5 + 10 = 15
   10 + 5 = 15
   15 – 5 = 10
   15 – 10 = 5
7. All of the addends in each problem make 15.

Challenge: Answers will vary.

page 38

1. Already done
2. 5 tens, 7 ones, 57
3. 9 tens, 3 ones, 93
4. 5 tens, 0 ones, 50
5. 6 tens, 5 ones, 65
6. 2 tens, 1 one, 21
7. 4 tens, 9 ones, 49
8. 2 tens, 2 ones, 22
9. 6 tens, 8 ones, 68
10. 16, 21, 22, 49, 50, 57, 65, 68, 93

page 39

1. 69
2. 50
3. 59
4. 88
5. 89
6. 95
7. 90
8. 49
9. 66
10. 58
11. 95
12. 98
13. 80

14. 97
15. 28
16. 99
17. 99
18. 60
19. 62
20. 21
21. 21
22. 10
23. 30
24. 1
25. 0
26. 31
27. 5
28. 5
29. 34
30. 30
31. 21
32. 15
33. 40
34. 62
35. 32
36. 0

page 40

Answers will vary.

page 41

6 different combos:

c-v-s	v-c-s
c-s-v	s-c-v
v-s-c	s-v-c

pages 42 and 43

Answers will vary.

page 44

R-G	O-P
R-B	Y-G
R-P	Y-B
O-G	Y-P
O-B	9 outfits